FLORA OF TROPICAL EAST AFRICA

FLAGELLARIACEAE

D. M. NAPPER

Shrubs or lianes. Leaves alternate, with split or entire often auricled sheathing bases; blade flat or plicate, with a long acuminate straight or cirriform tip; venation fine, parallel. Inflorescence a panicle. Flowers bracteate, hermaphrodite or dioecious, regular. Perianth subpetaloid or glumaceous, of 6 segments in 2 whorls. Stamens 6, opposite the perianth-segments; anthers basifixed, dithecous. Ovary superior, globose, 3-locular, with a short style and 3 linear stigmas. Fruit a small drupe, 3–1-seeded. Seed hemispheric or spherical.

A small family of 3 genera occurring in the tropical forests of Asia, Australasia and Africa. One genus and two species have been recorded from mainland Africa, though *F. indica* L. seems to be extremely rare.

FLAGELLARIA

L., Sp. Pl.: 333 (1753) & Gen. Pl., ed. 5: 156 (1754)

Scandent lianes, with terete smooth stems up to 2·5 cm. in diameter and completely covered by the leaf-sheaths when young. Leaf-sheath parallel-veined, slit or entire; leaf-blade abruptly narrowed at the base, usually with a short petiolar or subpetiolar base and a long attenuate cirriform tip. Flowers hermaphrodite, mostly crowded towards the tips of the branches. Perianth-segments subpetaloid. Stamens as many as the perianth-segments, some sometimes reduced to staminodes. Style long, divided almost to the base into 3 (or 2) linear stigmas. Drupe small, subglobose or ovoid, with a woody endocarp, 1-locular, 1-seeded, rarely 2-locular and 2-seeded. Seed globose with a membranous testa.

Leaf-sheath slit for at least ⅓ of its length, margins so
    tightly overlapped that the sheath often looks entire
    and notched above; flowers in short distinct racemes   1. *F. guineënsis*
Leaf-sheath entire, rarely torn, subtruncate at the mouth,
    with a chartaceous margin opposite the blade; flowers
    congested in very short racemes or glomerules .   .   2. *F. indica*

1. **F. guineënsis** *Schumach.* in Schumach. & Thonn., Beskr. Guin. Pl.: 181 (1827); Hiern in Cat. Afr. Pl. Welw. 2: 81 (1899); T.T.C.L.: 237 (1949); U.O.P.Z.: 266, fig. (1949); F.W.T.A., ed. 2, 3: 51 (1968). Type: Ghana [Guinea], *Thonning* (C, holo., K, microfiche!)

Scandent climber to 5(–10) m. high, climbing by means of the involute leaf-tip tendrils. Leaf-sheaths cylindric, slit to the base or almost so, with inconspicuously overlapped scarious margins, the shoulders rounded or subauriculate with scarious margins; leaf-blade oblong-lanceolate, with or without an abruptly widening subpetiolar base, and with a cirriform involute tip, 12–18 cm. long excluding the tendril, (1·4–)1·8–2·4(–2·8) cm. wide, entire

1

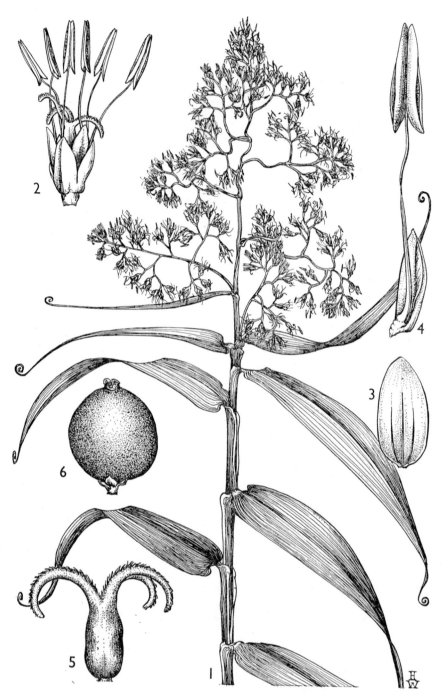

FIG. 1. *FLAGELLARIA GUINEENSIS*—**1**, part of shoot showing leaf-sheaths and inflorescence, × ⅔; **2**, flower, × 5; **3**, perianth-segment, × 10; **4**, stamen, with attached perianth-segment, × 10; **5**, gynoecium, × 10; **6**, fruit, × 4. 1, from *Tanner* 3736; 2–6, from *Faulkner* 1609.

at the margins, glabrous.   Inflorescence a terminal panicle, 8–12 cm. long, with flowers subsessile in short racemes; rhachis and branches glabrous; bracts at the base of the pedicel ovate, 1–1·25 mm. long, membranous; pedicel about as long as the bract.   Perianth-segments 6, rarely 8, ovate, 2–3 mm. long in subequal whorls, those of the outer whorl slightly smaller, white or cream.   Stamens exserted, as many as the perianth-segments; filaments attached at the base to the perianth, the alternate series sometimes reduced to short naked filaments.   Ovary globose, with a terminal style and 3 (rarely 2) linear stigmas.   Drupe spherical, 5–9 mm. in diameter, red, with persistent style-base and perianth.   Fig. 1.

KENYA.   Kwale District: Shimba Hills, Mwele Mdogo Forest, 6 Feb. 1953, *Drummond & Hemsley* 1152!; Mombasa I., Feb. 1876, *Hildebrandt* 1047b!; Kilifi District: Malindi, Sabaki R., Oct. 1965, *Tweedie* 3139!
TANGANYIKA.   Morogoro District: Turiani Forest, 24 Mar. 1954, *Padwa* 309!; Ulanga District: Kisawasawa, 19 Aug. 1959, *Haerdi* 314/0! Lindi District: Lake Lutamba, 23 Nov. 1966, *Gillett* 18012!
ZANZIBAR.   Zanzibar I., Mbiji, 2 Dec. 1930, *Greenway* 2634! & Mazizini [Massazine], 6 Feb. 1960, *Faulkner* 2485!; Pemba, Pembe I., 11 Oct. 1929, *Vaughan* 707!
DISTR.   **K**7; **T**3, 6, 8; **Z**; **P**; West Africa, Congo, Mozambique, Madagascar and South Africa
HAB.   Forest fringes and thicket; 0–400 m.

SYN.   [*F. indica* sensu Engl., P.O.A. C: 133 (1895); T.T.C.L.: 237 (1949), pro parte, non L.]
   F. indica L. var. *guineënsis* Engl. in V.E. 2: 257, fig. 174 (1908); Marloth, Fl. S. Afr. 4: 56, fig. 11 (1915)

2. **F. indica** *L.*, Sp. Pl.: 333 (1753); T.T.C.L.: 237 (1949); Backer in Fl. Males., ser. 1, 4: 247 (1951).   Type: without locality or collector, Herb. Linn. 463. 1 (LINN, lecto., K, microfiche!)

Perennial climber to 5 m. high.   Leaf-sheaths cylindric, entire, longi-tudinally ribbed, subtruncate, with a chartaceous rim on the margin opposed to the blade; leaf-blade broadly lanceolate or linear, rounded below and abruptly contracted into a subpetiolar base 3–10 mm. long, attenuate above, with a cirriform involute tip, 14–18 cm. long, 15–20 mm. wide, entire at the margins, glabrous.   Inflorescence a terminal panicle, with a stout bilaterally compressed rhachis and branches; flowers in subsessile very short subglobose spikes; rhachis and branches glabrous; bracts ovate, 1–1·25 mm. long, membranous.   Perianth-segments 6, rarely 8, ovate, 2–3 mm. long in 2 subequal whorls, those of the outer whorl slightly smaller, white or greenish. Stamens exserted, equalling the perianth-segments in number.   Ovary globose, with a terminal style and 3 linear stigmas.   Drupe spherical, 5–6 mm. in diameter, red.

TANGANYIKA.   Lindi District: Rondo Plateau, 12 June 1906, *Braun in Herb. Amani.* 1250!
DISTR.   **T**8; Indo-Malesia, Mozambique and the Mascarene Is.
HAB.   Forest and thicket; 0–400 m.

INDEX TO FLAGELLARIACEAE

Flagellaria *L.*, 1
Flagellaria guineënsis *Schumach.*, 1
Flagellaria indica *L.*, 3
  var. *guineënsis* Engl., 3
Flagellaria indica sensu auct., 3